FLORA

OF

TROPICAL EAST AFRICA

prepared at the Royal Botanic Gardens, Kew in cooperation with the East African Herbarium, the National Herbarium of Tanzania and the Herbaria of Makerere University and Dar es Salaam University

EDITOR

H.J. BEENTJE, M.Sc., Ph.D., F.L.S.

SUB-EDITOR

S.A. GHAZANFAR, M.Sc., Ph.D

HYDNORACEAE

BY

HENK BEENTJE
&
QUENTIN LUKE

CRC Press
Taylor & Francis Group
Boca Raton London New York

CRC Press is an imprint of the
Taylor & Francis Group, an **informa** business
A BALKEMA BOOK

LIST OF ABBREVIATIONS

A.V.P. = O. Hedberg, Afroalpine Vascular Plants; **B.J.B.B.** = Bulletin du Jardin Botanique de l'Etat, Bruxelles; Bulletin du Jardin Botanique Nationale de Belgique; **B.S.B.B.** = Bulletin de la Société Royale de Botanique de Belgique; **C.F.A.** = Conspectus Florae Angolensis; **E.J.** = A. Engler, Botanische Jahrbücher für Systematik, Pflanzengeschichte und Pflanzengeographie; **E.M.** = A. Engler, Monographieen Afrikanischer Pflanzen-Familien und Gattungen; **E.P.** = A. Engler, Das Pflanzenreich; **E.P.A.** = G. Cufodontis, Enumeratio Plantarum Aethiopiae Spermatophyta; in B.J.B.B. 23, Suppl. (1953) et seq.; **E. & P. Pf.** = A. Engler & K. Prantl, Die Natürlichen Pflanzenfamilien; **F.A.C.** = Flore d'Afrique Centrale (*formerly* F.C.B.); **F.C.B.** = Flore du Congo Belge et du Ruanda-Urundi; Flore du Congo, du Rwanda et du Burundi; **F.D.-O.A.** = A. Peter, Flora von Deutsch-Ostafrika; **F.F.N.R.** = F. White, Forest Flora of Northern Rhodesia; **F.P.N.A.** = W. Robyns, Flores des Spermatophytes du Parc National Albert; **F.P.S.** = F.W. Andrews, Flowering Plants of the Anglo-Egyptian Sudan *or* Flowering Plants of the Sudan; **F.P.U.** = E. Lind & A. Tallantire, Some Common Flowering Plants of Uganda; **F.R.** = F. Fedde, Repertorium Speciorum Novarum Regni Vegetabilis; **F.S.A.** = Flora of Southern Africa; **F.T.A.** = Flora of Tropical Africa; **F.W.T.A.** = Flora of West Tropical Africa; **F.Z.** = Flora Zambesiaca; **G.F.P.** = J. Hutchinson, The Genera of Flowering Plants; **G.P.** = G. Bentham & J.D. Hooker, Genera Plantarum; **G.T.** = D.M. Napper, Grasses of Tanganyika; **I.G.U.** = K.W. Harker & D.M. Napper, An Illustrated Guide to the Grasses of Uganda; **I.T.U.** = W.J. Eggeling, Indigenous Trees of the Uganda Protectorate; **J.B.** = Journal of Botany; **J.L.S.** = Journal of the Linnean Society of London, Botany; **K.B.** = Kew Bulletin, *or* Bulletin of Miscellaneous Information, Kew; **K.T.S.** = I. Dale & P.J. Greenway, Kenya Trees and Shrubs; **K.T.S.L.** = H.J. Beentje, Kenya Trees, Shrubs and Lianas; **L.T.A.** = E.G. Baker, Leguminosae of Tropical Africa; **N.B.G.B.** = Notizblatt des Botanischen Gartens und Museums zu Berlin-Dahlem; **P.O.A.** = A. Engler, Die Pflanzenwelt Ost-Afrikas und der Nachbargebiete; **R.K.G.** = A.V. Bogdan, A Revised List of Kenya Grasses; **T.S.K.** = E. Battiscombe, Trees and Shrubs of Kenya Colony; **T.T.C.L.** = J.P.M. Brenan, Check-lists of the Forest Trees and Shrubs of the British Empire no. 5, part II, Tanganyika Territory; **U.K.W.F.** = A.D.Q. Agnew (or for ed. 2, A.D.Q. Agnew & S. Agnew), Upland Kenya Wild Flowers; **U.O.P.Z.** = R.O. Williams, Useful and Ornamental Plants in Zanzibar and Pemba; **V.E.** = A. Engler & O. Drude, Die Vegetation der Erde, IX, Pflanzenwelt Afrikas; **W.F.K.** = A.J. Jex-Blake, Some Wild Flowers of Kenya; **Z.A.E.** = Wissenschaftliche Ergebnisse der Deutschen Zentral-Afrika-Expedition 1907–1908, 2 (Botanik).

FAMILIES OF VASCULAR PLANTS REPRESENTED IN THE FLORA OF TROPICAL EAST AFRICA

The family system used in the Flora has diverged in some respects from that now in use at Kew and the herbaria in East Africa. The accepted family name of a synonym or alternative is indicated by the word "see". Included family names are referred to the one used in the Flora by "in" if in accordance with the current system, and "as" if not. Where two families are included in one fascicle the subsidiary family is referred to the main family by "with".

Published by:
CRC Press/Balkema
P.O. Box 447, 2300 AK Leiden, The Netherlands
e-mail: Pub.NL@taylorandfrancis.com
www.crcpress.com – www.taylorandfrancis.com

CRC Press/Balkema is an imprint of the Taylor & Francis Group, an informa business

ISBN 13: 978-90-5809-408-7 (pbk)

FLORA OF TROPICAL EAST AFRICA

HYDNORACEAE

HENK BEENTJE & QUENTIN LUKE

Mostly subterranean, leafless, plants without chlorophyll; body rhizome-like with many slender, unbranched, haustorial roots, parasitic on the roots of the host plant. Flowers solitary, emerging from the rhizome-like body, hermaphrodite, the lower part subterranean, large, fleshy, with strong smell, beetle-pollinated; perianth tubular, lobed at apex, lobes 3–4(–5), valvate, fleshy; anthers sessile or very shortly stalked, 3–4(–5), inserted on the hypanthium opposite the perianth lobes, extrorse, with elongate many-locular pollen sacs with slits, forming a lobed ring on the hypanthium (*Hydnora*), or connate anthers forming a dome (*Prosopanche*), with (*Prosopanche*) or without staminodes (*Hydnora*). Carpels 3(–4), ovary 1-locular, inferior, becoming filled with accrescent placentas covered with many ovules; ovules anatropous, unitegmic; stigma sessile, 1–5-lobed. Fruit subterranean, fleshy, with woody pericarp; seeds many, minute.

Two genera, *Hydnora* in Africa and Madagascar, *Prosopanche* De Bary in Central and S America.

HYDNORA

Thunb. in Kongl. Svenska Vetenskapakad. Handl. 36: 69, t. 2 (1775)

Leafless, chlorophyll-less, root parasites; plant body subterranean, large, warty, rhizome-like structure with many branches. Flowers arising directly from the plant body, partly subterranean. Anthers sessile, forming a lobed ring on the hypanthium, with elongate, bisporangiate pollen sacs with longitudinal slits. Stigma mostly 3–5-lobed. Fruit a subterranean fleshy berry with a woody pericarp; seeds many.

3–7 species in tropical and South Africa and Madagascar. In older literature such as in F.T.A. and Index Kewensis often included in the family *Cytinaceae*.

Flower lobes with the margins in their proximal part beset by a broad band of setae, the apex of different texture, more fleshy (?horny), glabrous .... 1. *H. abyssinica*
Flower lobes margins glabrous and homogenous .... 2. *H. sinandevu*

1. **H. abyssinica** *A. Braun* in Schweinf., Beitr. Fl. Aethiop.: 217 (1867); Engl. in P.O.A. C: 169 (1895); Baker & C.H. Wright in F.T.A. 6(1): 133 (1913); F.P.S. 1: 22 (1950); Blundell, Wild Flow. E. Afr.: photo 541 (1987); Musselman in F.Z. 9(2): 18, t. 7 (1997). Type: Ethiopia, mountains near Dehli-Dikeno, *Schimper* 963 (B, holo., P!, ?W, iso.)

Subterranean root parasite. Plant body ('branch') rhizome-like, simple or with widely spreading horizontal branches, fleshy, irregularly shaped, to 1 m long, to 10 cm wide, somewhat flattened, brown outside, brick-red or reddish pink to white inside, with sticky, astringent exudate when fresh, side branches cylindrical or nearly so,

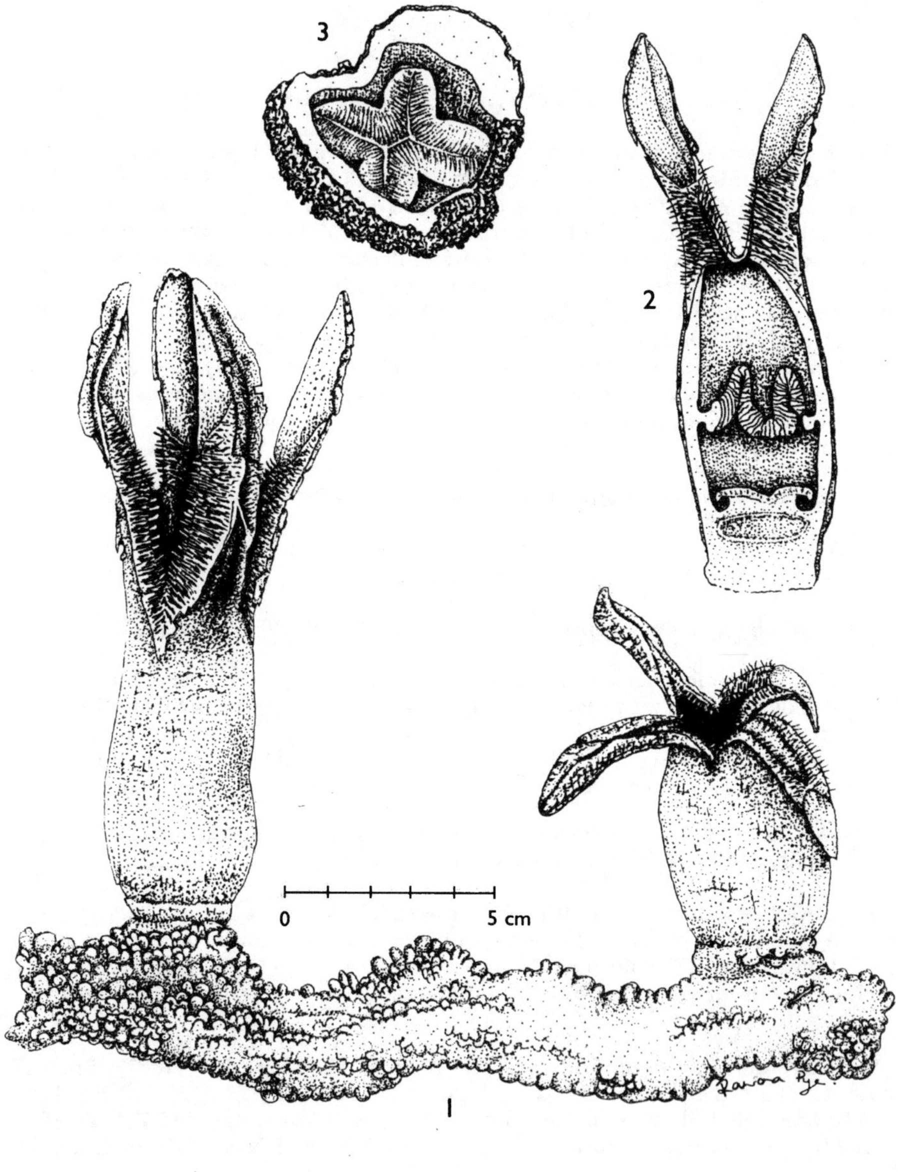

FIG. 1. *HYDNORA ABYSSINICA* — **1**, habit; **2**, longitudinal section of flower; **3**, cross-section of flower. All from *Luke & Luke* 7182. Drawn by Rawana Pye.

1.7–4 cm in diameter, densely covered with small warts and with flower buds on the upper surface. Flowers emerging directly from the main and side branches, remaining partially below ground and partially emerging, solitary or in groups, but only one per branch emergence point (?), with very foetid smell; pedicel absent or up to 2.5 cm long. Perianth fleshy-coriaceous, (8?–)11–25(–33) cm long; tube 4-sided to terete, reddish brown and smooth outside, pinkish to ivory white inside, 4.8–14(–19) cm long (depending on depth of rhizome), 3–5.5 cm in diameter; lobes (2?–3–)4(–6), emergent from ground, either (?in dry weather or in early stage) erect and connivent by the thickened tips with only side vents open, or (?in moist weather or at a later stage) spread and flat on ground, lanceolate to spatulate, thickly fleshy, 4–9(–14) cm long, 1.7–3.8 cm wide, broadly channelled on the inside, lower $^1/_3$ to $^2/_3$ of margins bright red, turning pink or cream/yellow, to 10–13 mm wide, densely setose with broad-based tapering white setae 1–7 mm long, lobe tips ('cucullus') thickened, triangular in cross-section, apricot or cream to white, 2.6–5 cm long and 1.6–2 cm wide, grooved outside, the two inside facets glabrous and cushion-like; stamens as many as the lobes, with 2 large anthers, 2.5–3.4 cm long, 2–2.5 cm wide, anthers inserted 5–8(–10.5) cm above the base of tube (approximately mid-way), continuously joined at their bases in a series of inverted V-shapes forming a wavy cream, yellow or orange ring with the top (30–)43–102 mm above the flower base, this transversely striate and divided into numerous horizontal pollen sacs; pollen sticky, adhering to the anthers; ovary unilocular, subglobose, 22–28(–45) mm in diameter with numerous infolded patent placentas; stigma sessile, cushion-shaped, pink, 3–4(–5)-lobed, each lobe finely ribbed and covered with clear or brown syrupy liquid when fresh. Fruit entirely subterranean, fleshy, globose, 10–15 cm in diameter, many-seeded, often splitting irregularly at maturity; outer periderm scaly; inner pericarp mealy, white; seeds brown, oblong to globose, 1–1.8 mm long. Figs. 1 & 2. (pp. 2 & 4).

UGANDA. Karamoja District: Lochoi, May 1940, *A.S. Thomas* 3539! & Amudat, May 1948, *Eggeling* 5812!; Ankole District: Ruizi R., Feb. 1951, *Jarrett* 444!

KENYA. Northern Frontier District: 30 km on Ramu–Malka Mari road, May 1978, *Gilbert & Thulin* 1534!; Nairobi: Nairobi National park, just outside Cheetah Gate, Apr. 1975, *Friis & Hansen* 2605!; Kajiado District: Chyulu Hills, Oct. 1969, *Gillett & Kariuki* 18797!

TANZANIA. Mbulu District: Lake Manyara National Park, Msasa, Dec. 1963, *Greenway & Kirrika* 11132!; Mpanda District: Katisunga, Jan. 1950, *Bullock* 2273!; Kondoa District: Kondoa, Great North Road, Jan. 1962, *Polhill & Paulo* 1139!

DISTR. **U** 1, 2, 4; **K** 1–4, 6, 7; **T** 1–5; Congo (Kinshasa), Rwanda, Sudan, Eritrea, Ethiopia, Somalia, Zimbabwe, Botswana, Namibia, Swaziland, South Africa; Yemen

HAB. In dry woodland, wooded grassland or bushland, on roots of *Acacia* (various species), *Albizia, Adansonia, Delonix, Kigelia, Piliostigma*; 90–2100 m

USES. Branch/'rhizome' rich in tannins; flower buds eaten by Pokot (fide *Timberlake*) and Maasai ("Orukunyi" in KiMaa). Inner pericarp of fruit sweet to taste, edible, also eaten by rhino, wild pig, warthog; minor medicinal use for childbirth.

CONSERVATION STATUS. Least Concern (LC)

SYN. *H. johannis* Becc. in Nuov. Giorn. Bot. Ital. 3: 5 (1871); Solms in E.P. 4, fam. 76: 7 (1901); Harms in E. & P. Pf. ed. 2, 16b: 291 (1935); Lebrun in F.C.B. 1: 392 (1948); Malaisse in B.J.B.B. 52: 115, fig. 1 (1982); Musselman in Notes Roy. Bot. Gard. Edin. 42: 23, fig. (1984) & in Dinteria 19: 77 (1987) & in Thulin, Fl. Somalia 1: 31, fig. 16 (1993). Type: Eritrea, Barka Valley, Keren–Bogos, also Sciotel 1870, *Beccari* 417, 423, 424, 3685, 3687, 3690 (FT!, syn.)

H. bogosensis Becc. in Nuov. Giorn. Bot. Ital. 3: 6 (1871); Baker & C.H. Wright in F.T.A. 6(1): 134 (1913). Type: Eritrea, not specified, Keren, fl. Jul.–Aug. 1870, *Beccari* s.n. (FT!, holo.)

H. aethiopica Decne. in Bull. Soc. Bot. Fr. 20: 77 (1873); Baker & C.H. Wright in F.T.A. 6(1): 132 (1913). Type: Ethiopia, *Sabatier* s.n. (P!, holo.)

H. hanningtonii Rendle in J.B. 34: 55, t. 356 (1896); Baker & C.H. Wright in F.T.A. 6(1): 133 (1913). Syntypes: ?Uganda/Kenya, Victoria Nyanza, Jordan's nullah, Dec. 1882, *Bishop Hannington* s.n. & Ethiopia, Galla Highlands, Ginia, 1895, *Donaldson Smith* s.n. (both BM!, syn.)

H. abyssinica A. Braun var. *quinquefida* Engl. in E.J. 28: 386 (1900); Baker & C.H. Wright in F.T.A. 6(1): 134 (1913). Type: Tanzania, Uhehe, Lukosse R., *Goetze* 487 (B, holo.)

FIG. 2. *HYDNORA ABYSSINICA* — Kenya, Kajiado District, Chyulu Hills below Bonham House, 22 Dec. 2000, *Luke & Luke* 7182. Photograph by Quentin Luke.

H. johannis Becc. var. *quinquefida* (Engl.) Solms in E.P. 4, 76: 7 (1901)
H. ruspolii Chiov. in Boll. Soc. Bot. Ital. 1917: 57 (1915). Type: Ethiopia, Ogaden, Marro Umberto, *Riva & Ruspoli* 1091 (FT!, holo.)
H. gigantea Chiov., Result. Sci. Miss. Stef.-Paoli, Coll. Bot. 1: 156 (1916). Type: Somalia, Berdale to El Ualac, 1911, *Paoli* 980 & 1035 (FT, holo., not found)
H. gigantea Chiov. var. *trimera* Chiov., Result. Sci. Miss. Stef.-Paoli, Coll. Bot. 1: 157 (1916). Types: as *H. gigantea*
H. michaelis Peter in F.D.-O.A. 2: 185, t. 21 (1932). Type: Tanzania, Usagara, Kidete, Pori on Romuma (sic) R., *Peter* 32690 (B, holo.)
H. cornii Vacc. in Nuov. Giorn. Bot. Ital., n.s. 39: 304, fig. 1, t. 14 (1932). Type: Somalia, Villa Governationale di Afgoi, Mogadishu, ?*S.E. Corni* s.n. (TOM, holo.)

NOTE. The species has recently been cited as *A. Braun ex Schweinf.* or *ex Decne.* but the family was treated by Braun. The description is short, but valid, and therefore this name has priority over the often-used synonym *H. johannis*.

The flowers emerge just before or at the start of the rains and last for a few days; flowering in one area lasting for about a month.

All flowers seen by us are ± sessile; they are said to be stalked for 2.5–3 cm (according to Chiovenda in the protologue of *H. gigantea*) or 5–9 cm long (fide Musselman, L.J., 1984. Some parasitic angiosperms of Sudan: Hydnoraceae, Orobanchaceae, and Cuscuta (Convolvulaceae). Notes Roy. Bot. Gard. Edin., 42: 21–38).

Musselman states the thickened lobe tips are the scent-producing area.

Flowers contain beetles of genus *Geotrupes* (dung beetles) and 'flies' fide *Burtt* 474; maggots and two kinds of beetles of the family Tenebrionidae, black *Rhytinota* sp. and grey *Psammodes* sp. were found in *Luke & Luke* 7182.

A specimen from Uganda, **U** 4, Mengo District: Bulemezi, Sekamuli, June 1967, *Ferreira* NC133! is very short and has the basal spherical swelling that is more characteristic of *H. africana* Thunb. It is recorded to grow on *Combretum*, and though the collector says 'no hairs' there are a few present. More material is needed to be certain of this taxon.

2. **H. sinandevu** Beentje & Q. Luke **sp. nov.** *H. abyssinicae* A. Braun affinis sed corollae lobis glabris (non setulosis) atque absentia corporis callosi ab ea differt; ramis subterraneis verrucosis in lineis (non utrinque verrucosis). Typus: Kenya, Kilifi District, Arabuko-Sokoke, Mida, Jan. 1992, *Luke* 3033a (EA!, holo., K!, MO!, US!, iso.)

Subterranean root parasite. Plant body ('branch') not fully known known, parts seen to 2 cm in diameter, warty in lines, warts to 4 mm across. Flowers remaining partially below ground and partially emerging; pedicel to 8 mm long. Perianth coriaceous, 8–16 cm long; tube brown outside, reddish or pale pink inside, 5.3–9 cm long, 1.9–2.8 cm in diameter; lobes 4, emergent from ground, yellowish brown (outside?) to red or yellow inside, narrowly lanceolate, 2.7–6.5 cm long, 0.9–2.9 cm wide, not setose on margins, lobe tips 3.5–5 cm long and 2–2.2 cm wide, ?flat outside, the two inside facets glabrous, ?brown; stamens inserted 22–50 mm above base of tube; anthers continuously joined at their bases in a series of inverted V-shapes forming a wavy ring 15–20 mm high with the top 46–73 mm above the flower base, this transversely striate and divided into numerous horizontal pollen sacs; pollen sticky, adhering to the anthers; ovary unilocular, subglobose, 15–30 mm (when dry) in diameter with numerous infolded patent placentas. Fruit not seen. Fig. 3 (p. 6).

KENYA. Kwale District: Taru Desert, Oct. 1956, *Bally* 10434!; Tana R District: S of Garissa, Jan. 1943, *Bally* 2051!; Kilifi District: Arabuko-Sokoke, Mida, Jan. 1992, *Luke* 3033a!

TANZANIA. Lushoto District: W Usambaras, Daluni, Aug. 1935, *Davies* 1004!; Pangani District: Msubugwe S at Pangani R., Mar. 1950, *Verdcourt* AH9924!; ?Bagamoyo District: mainland opposite Zanzibar, 1884, *Kirk* 150!

DISTR. **K** 1, 7; **T** 2? (see note below), 3, 6?; not known elsewhere

HAB. In scattered tree grassland, *Acacia–Commiphora* bushland, thicket or forest margin between mangrove and forest, on *Commiphora campestris* or *C. africana* roots (*Luke et al.* TPR677, 3033a), possibly on *Pterocarpus* or *Ostryoderris* (fide *Verdcourt*); 2–450 m

USES. Minor medicinal use against throat complaints

CONSERVATION STATUS. Data Deficient (DD)

NOTE. A specimen from **T** 2, Masai District: Olduvai Gorge, Apr. 1964, *Mrs L.S.B. Leakey* H97/64! is filled with hundreds of small beetles; it has a pedicel of 13 mm, the perianth is 19 cm long, the tube is 11.5 cm long and 4.5 cm in diameter; lobes are four, 7 × 2.2 cm; ovary 3.8 cm high, anthers inserted at 68 mm with their apex at 95 mm. This is most likely this taxon, but the altitude is about 1500 m.

K 1, Northern Frontier District: Kailongol Mts, 1°50' N 35°50' E, June 1970, *Mathew* 6850! is probably this taxon as well, but the material is insufficient to be certain.

FIG. 3. *HYDNORA SINANDEVU* — Kenya, Kilifi Distribt, Arabuko-Sokoke, Mida Creek, Jan. 1992, *Luke* 3033a. Photograph by Quentin Luke.

INCERTAE SEDIS

Several specimens are too young, or too fragmented to be identified. These include: *Kanywa* 84 from **T** 1, Bukoba; *Joanna* CM8635 from **K** 7, Kasigau; **T** 2, *BLI Amani* 629, Bezirk Moshi am Kilimandjaro, Nov. 1904, Dom..u (Pflanzer) (illegible!).

INDEX TO HYDNORACEAE

New names validated in this part

Hydnora sinandevu *Beentje & Q. Luke* **sp. nov.**

GEOGRAPHICAL DIVISIONS OF THE FLORA

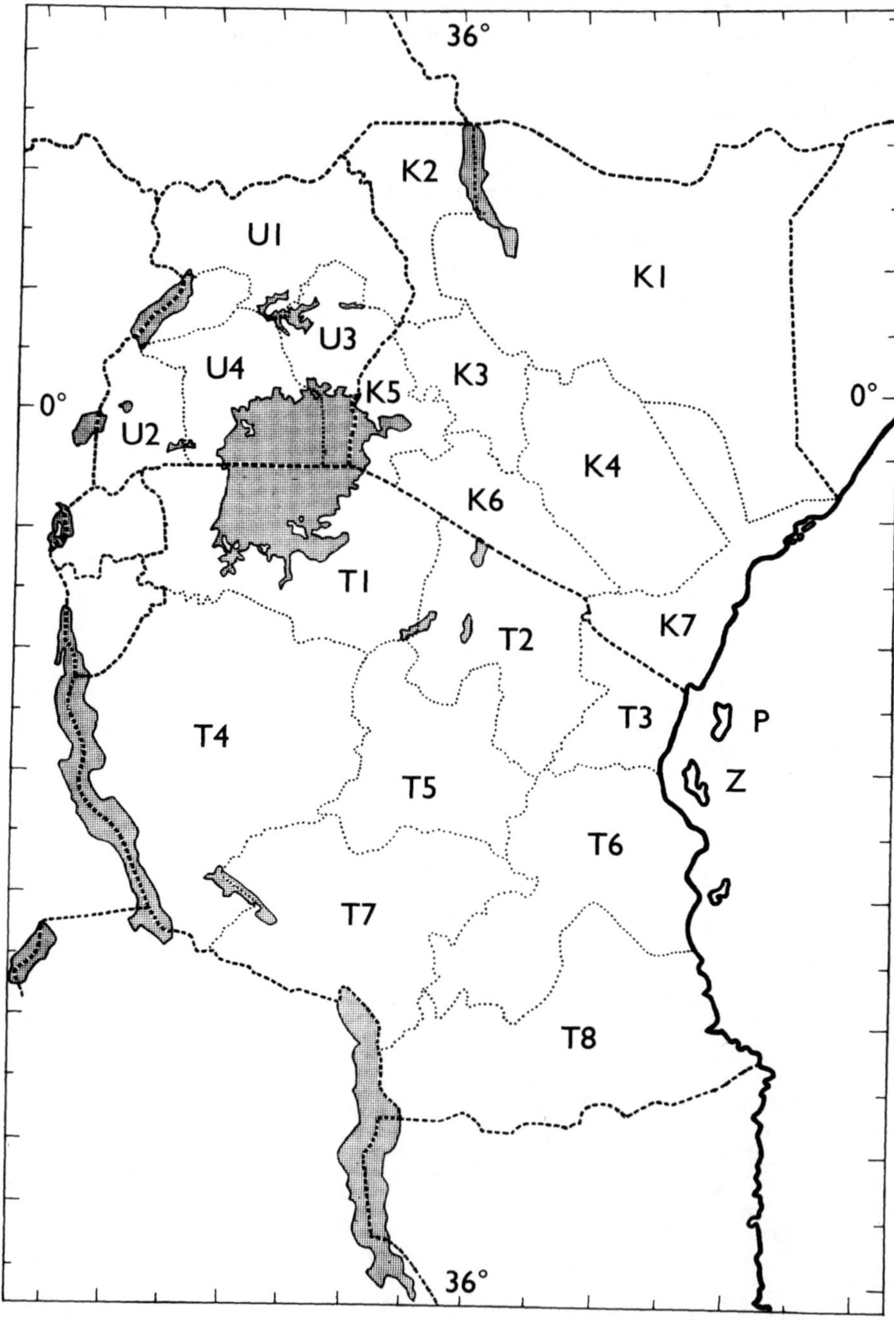